How Does It Grow?

HOW DO BANANAS GROW?

By Kathleen Connors

Gareth Stevens
PUBLISHING

Please visit our website, www.garethstevens.com. For a free color catalog of all our high-quality books, call toll free 1-800-542-2595 or fax 1-877-542-2596.

Library of Congress Cataloging-in-Publication Data

Names: Connors, Kathleen, author.
Title: How do bananas grow? / Kathleen Connors.
Description: New York : Gareth Stevens Publishing, [2022] | Series: How does it grow? | Includes index.
Identifiers: LCCN 2020011774 | ISBN 9781538262955 (library binding) | ISBN 9781538262931 (paperback) | ISBN 9781538262948 (6 Pack) | ISBN 9781538262962 (ebook)
Subjects: LCSH: Bananas–Juvenile literature. | Bananas–Life cycles–Juvenile literature.
Classification: LCC SB379.B2 C66 2022 | DDC 634/.772–dc23
LC record available at https://lccn.loc.gov/2020011774

First Edition

Published in 2022 by
Gareth Stevens Publishing
111 East 14th Street, Suite 349
New York, NY 10003

Editor: Kristen Nelson
Designer: Katelyn E. Reynolds

Photo credits: Cover, p. 1 pp1/Shutterstock.com; p. 5 bbostjan/E+/Getty Images; p. 7 Kriangkrai Thitimakorn/ Moment / Getty Images Plus; p. 9 Gheorhge/ iStock / Getty Images Plus; p. 11 amnachphoto/ iStock / Getty Images Plus; pp. 13, 24 (leaves) Thissatan/ iStock / Getty Images Plus; pp. 15, 24 (flowers) Uros Ravbar/ Lonely Planet Images / Getty Images Plus; p. 17 Wolfgang Kaehler/LightRocket via Getty Images; pp. 19, 24 (harvest) Alex Craig/ Photolibrary / Getty Images Plus; p. 21 Hispanolistic/E+/Getty Images; p. 23 bhofack2/ iStock / Getty Images Plus.

Printed in the United States of America

CPSIA compliance information: Batch #CSGS22: For further information contact Gareth Stevens, New York, New York at 1-800-542-2595.

Contents

Bananas are yummy!
How do they grow?

They grow in warm places.
They grow on farms.

They grow from part
of another banana plant.

They are planted in soil.
They need lots of water.

Roots and leaves grow.
The leaves grow together.
They look like a tree trunk!

The plants grow very tall.
They grow flowers.

Some become bananas.
They are green at first.

They are cut
from the plant.
This is the harvest.

Most bananas turn yellow.
They are ready to eat.

Red bananas have
red skin.
They taste sweet.

Words to Know

flower

harvest

leaves

Index